Foreward

I grew up in a small town in Ohio. I was the youngest. My older brothers and sister had a chemistry laboratory in the basement. I yearned to get in to do experiments but my older brothers restricted my access. They had serious work to do there. When they went off to college, I finally had free access to the lab and all of its mysterious powders, test tubes, flasks, chemistry manuals– and a big radio which our father brought home from the USS Intrepid, the Navy aircraft carrier where he served. I imagined myself as a secret scientist creating exotic mixtures with profound properties.

In this book, wonderfully illustrated and written by David Marshall, the main character Max is a boy whose curiosity leads him on a scientific path to solve a worldwide problem: Save the bees!

Max personifies my dream since I was that child in the basement laboratory. A great challenge we face today is the sudden loss of bees, our best pollinators for growing food. Bees are dying off in record numbers. The causes are complex. Unless we fix this, our global food security is at risk. Food could be become much more expensive. Ecosystems could unravel.

Max describes the challenge bees face and reveals a revolutionary insight to help bees. I hope this story inspires you to discover solutions to our problems. We need to protect this beautiful planet we call Earth, our home. We can do this! But we need young scientists like you to lead the way.

Join Max on this scientific adventure story.

Paul Stamets
Earthling

Invention Ambassador for the American Association for the Advancement of Science, 2014-2015.

"

Bees are so little,

but so important to us all.

The tiniest on Earth

can make the

biggest difference.

Paul Stamets

Night One

It was late spring, the day before Max's elementary school was getting out for summer vacation, when he learned about the trouble that the honey bees were in.

Max, sitting in his class, listened to Mr. Clark, the science teacher, talk about how more bees are dying than ever. The beekeepers began to notice that the bees were vanishing and leaving no clues as to why. Until now scientists didn't know the reason.

Mr. Clark talked about some of the causes that are hurting the bees. Pesticides, parasites, and pathogens are devastating the bee colonies. Mr. Clark explained that if something isn't done soon, within 10-20 years the honey bee population could be gone.

Max raised his hand, "What can we do to save them?" Mr. Clark just looked at Max and said, "I don't know, nobody knows how to save them."

The bell rang. The last bell signaling the start of summer! In a few days, Max would be heading out to visit his grandfather like he did every year.

As the bus rumbled home, Max daydreamed about his summers on the farm with his grandfather. He was getting excited.

Max got off the bus and ran into the house, "Mom!" Max yelled, excited he was out of school. He dropped his backpack and opened the refrigerator to grab a snack.

"Hi, Max! Are you looking forward to going to Grandpa's farm?" mom asked, seeing Max smile ear to ear.

"Oh, yes, mom! I am so excited! I can't wait to see Grandpa and help out on the farm! And this summer Grandpa said he would teach me how to fly Queen Bee."

Queen Bee is the biplane his grandfather found stored away in the back of an old hay barn. During its day it was used during WWII as a target plane. An operator, by remote control, guided the plane to fly back and forth out over the water as the war ships practiced firing at it. Queen Bee was very special. It never took one hit during the war, and so at the end of the war, it made its way to Ohio and was stored in the back of a hay barn. When Grandpa found it, he made a deal with the farmer who owned it. He offered the farmer free crop dusting services if he would allow him to take it. The farmer agreed, and Grandpa restored the old biplane back to its original condition, right down to the paint, black and yellow. Grandma thought it looked like a big bumble bee, so he named it Queen Bee, in honor of her.

Grandpa was so good at crop dusting that soon all the surrounding farmers hired him to dust their crops as well. Every year Max's grandfather helped all the farmers with their crops, spraying them with a special chemical that protected the crops from bugs and disease.

But that was years ago. With modern technology, the use of helicopters, and modern spraying equipment, the old biplane once again was retired. Grandpa would take Queen Bee out flying once in a while, waving as he flew over neighboring farmers. Most of the time it was stored in the back of the barn. Except this time instead of collecting dust and rotting away, Grandpa kept Queen Bee running in top shape. Now, every summer, Max, and his Grandpa would take the plane out and fly all day long – this passion for flying was something unique that Max and his grandpa shared.

Max loved his grandpa very much. They had a special relationship. To Max, there was nothing his grandpa couldn't fix or problem he couldn't solve. That seems to be the way it is with grandfathers.

For example, when the farmers were running out of power to operate their farming equipment and didn't know what to do, it was Grandpa who had the idea to build windmills. The farmers were so happy because the windmills produced enough energy to run all the farming equipment and had extra for the town to use also. Everyone in the town was amazed at the windmill's noiseless, clean and natural power. Now they didn't have to drill big holes in the ground to pull out oil or gas for energy.

At first, the farmers and the townspeople didn't see the benefits and didn't understand why anyone would want to build windmills. It became very clear, very quickly what the benefits were. Grandpa's idea proved to be one of the best ideas that the farmers could have hoped for.

Grandpa seemed to always have the answer for any challenge or problem.

Max's mother took him to the bus station early the next morning. Max was so excited. As the bus pulled up, Max gave his mother a big hug and a kiss. With the freshly baked cookies she made for him to take to grandpa, he stepped onto the bus and took his seat.

From the window, he looked out at his mother with a big smile, waved goodbye and sat in his seat full of anticipation.

A few hours later Max woke up as the bus was pulling into the bus station. He could read the sign, "Welcome to Columbiana, Ohio." Surprised, Max looked around and realized he had been dreaming. He got up and out of his seat, grabbed his suitcase and got off the bus. There was Grandpa, waiting for him with a big smile on his face.

Max dropped his bags and went running to Grandpa, and with a warm hug, Grandpa welcomed Max. "Welcome Max, so happy to see you!"

"I couldn't wait to see you, Grandpa!" Max replied.

"Maxie, we are going to have a memorable summer. I can just feel it," Grandpa replied as he picked up Max's suitcase and they headed for the truck.

Grandpa grew up in Columbiana, Ohio. It also happens to be where the first honey bee festival was established, called The Mesopotamia Honey Bee Festival. This is a big event, and Max looks forward to this festival every year.

"You still have the old truck, Grandpa?" Max asked with a grin.

"This truck is older than you Maxie. Your Great Grandpa bought this brand new, back in 1950. Some things you just can't get rid of Max, like old jeans, fishing rods, and this truck, right Maxie?" Grandpa always knew how to make his point. Max loved that old truck, and he knew what Grandpa meant by old jeans. They just feel more comfortable the older they get.

Driving home, Max looked out the window thinking about the farm and the plane. "So . . ." Max began, "how is Queen Bee doing?"

Smiling, grandpa knew the surprise waiting for Max. "Max you won't recognize her. She has a brand new paint job, and she's running like the day she was built."

"Wow, really Grandpa?" Max was dying to ask him if he remembered what he promised at the end of last summers' vacation. Max couldn't stand the anticipation, and just as he started to ask about flying Queen Bee, Grandpa said, "So, Maxie, what new things did you learn in school this year?" Grandpa knew perfectly well that Max didn't want to talk about school, only Queen Bee. Grandpa loved to kid around with Max, so Max, being a good boy, and respectful, began to talk about the honey bee.

"Grandpa, did you know that the honey bee is endangered? If something isn't done soon, within five years, they will be gone!"

"What? Where did you hear this?" Grandpa asked in disbelief.

"At school, my science teacher, Mr. Clark told us about how the bee farmers are going to their hives to check on them, and they discover that all the bees have vanished. Mr. Clark said that scientists have discovered that pesticides make them sick."

Surprised by all that Max knew about the danger that the honey bee is in, he asked Max to go on explaining about the bees.

"Yes, and Grandpa did you know that about 70% of our food relies on honey bees to pollinate it? If the bees go away, we won't have much of anything left to eat."

Wondering just how much Max knew about which crops would be affected, Grandpa asked, "Like what kind of food Max?"

"Oh, Grandpa you won't believe it!" Max began, "watermelon and strawberries, apples and cucumbers, onions and oranges, squash and peppers, I mean the list goes on and on. Mr. Clark said that 30% percent of our food is directly reliant on the honey bee and 70% of our food source is indirectly affected. Even the cows will suffer because the bees pollinate the clover and alfalfa, important crops that cows eat.

"Mr. Clark said that pathogens and parasites are also causing the devastation of the bees. You know, bees are very, very sensitive to their environment, Grandpa?"

Grandpa was very proud by how Max understood exactly how it would affect people and the planet.

"So how do they plan on saving the honey bee?" Grandpa asked.

"That's the problem they don't know how!" Max said feeling worried.

"Yes, I have heard, it's a very big problem," Grandpa replied. "Well, Max, you have been paying attention, haven't you? I am so proud of you. I have no doubt that one day you are going to do something great to benefit others."

As they drove down the highway towards Grandpa's farm, Max thought about the bees and wondered what he could do to help them. Daydreaming about the bees, Max forgot all about flying Queen Bee.

The drive home went by so fast. As they rounded the corner of the road, Max could see the farm house and the hay barn. He couldn't believe he was finally here. He had been looking forward to this all year.

As they pulled up to the house, Grandpa turned to Max, "You know Maxie, I was looking forward to you coming out this summer. The farm isn't the same without you."

Hearing Grandpa's voice always made Max feel that everything was alright in the world. It made him feel reassured.

Max jumped out of the truck, grabbed his suitcase and ran up to his room. Opening the door to his room, he ran and lept on his bed, feeling so happy to be out on the farm with Grandpa.

"Max!" Grandpa yelled, "time to come down for supper."

Max was in his room lying in his bed feeling a little tired from his trip, but the excitement about being with Grandpa on the farm seemed to wash away any tiredness he was feeling. Hearing his grandfather's voice calling to him, he got up and slid down the banister to the kitchen.

"Hey, what's for supper Grandpa, I'm starving?" Max asked.

"Well, we have fresh corn on the cob, tomatoes, squash, and peas from the garden. I also cooked us up some chicken. You like chicken right, Maxie?"

"Oh, yes, Grandpa," and as Max served the vegetables he reminded Grandpa, "You know, these vegetables won't exist anymore if anything happens to the bees. Do you think we could do something to help them?"

"Well, Maxie, I do believe that there is a solution to every challenge. If you and I put our heads together, who knows?" Grandpa replied, thinking about how serious it would be if the bees were gone.

After dinner, Max walked down to the hay barn to see Queen Bee. Walking up to the double winged plane Max thought, "Wow, to think I will learn to fly this beautiful plane this summer." Looking up at Queen Bee, he said, "If you could talk, I can only imagine the stories you would tell."

"Maxie!" came the voice of Grandpa calling and Max knew what that meant. It was time for bed. Max walked up to the house, and there was Grandpa on the porch. "Bedtime sure comes early on the farm Grandpa," Max replied, walking up the porch steps.

"That's because morning comes faster," Grandpa replied smiling. "And tomorrow we have a big day Max. I have a special surprise for you."

"What? What is it, Grandpa?" Max asked, so excited to hear what he had planned.

"Oh, you'll see Max, it's a surprise." And that is all he would say.

Max knew better than try to get Grandpa to tell him.

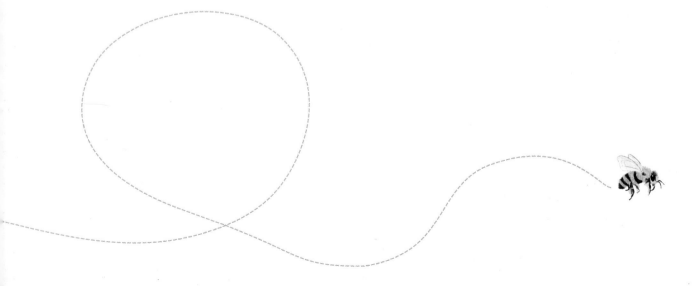

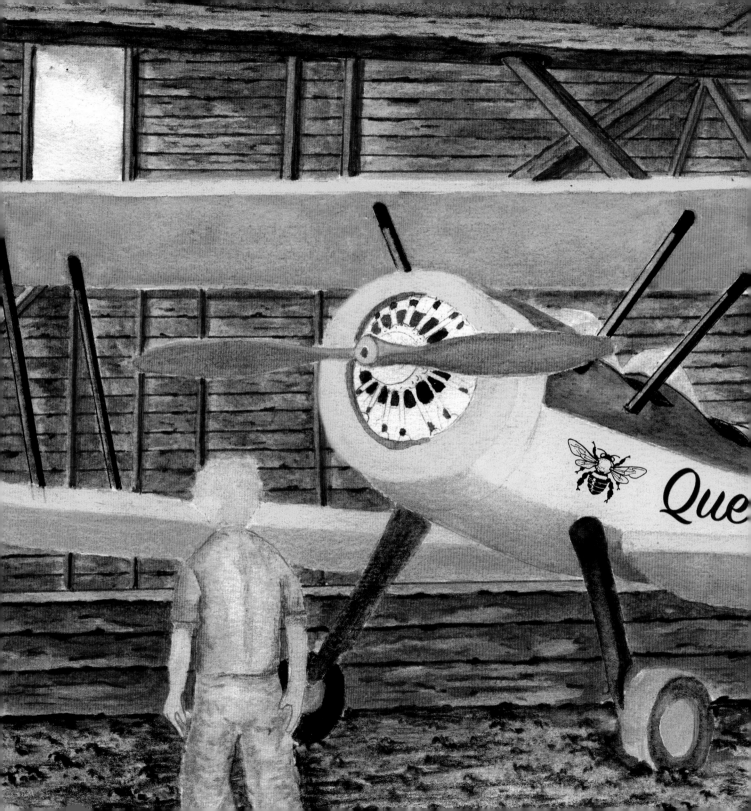

The next morning indeed came early, but Max was ready. He got dressed, tied his sneakers and ran down the stairs fully expecting to see Grandpa sitting at the kitchen table. A bowl of cereal and a glass of orange juice sat on the table with a note. Max picked up the note that read, "When you finish breakfast come down to the hay barn."

Max never ate so fast. He flew out the door, down the steps and ran to the hay barn. As he ran, he could see Queen Bee, the morning sun reflecting off her yellow wings. Grandpa rolled it out and had it ready to go.

"Maxie, today we are going flying!"

Max jumped with joy! Max loved flying in Queen Bee, up high with the clouds, and Grandpa. There was no feeling like it.

"Well, let's get started, Max, you know what to do," Grandpa said as he climbed into the cockpit of Queen Bee.

Max had been learning from Grandpa since he was small. He knew every inch of Queen Bee and had learned every step to getting her ready to fly.

Walking to the front of Queen Bee, Max stepped up to the propeller, moved it one way to open one valve, then turned it the other way to prime the engine. Max even knew the firing order of the pistons, and he began to call out, "1, 3, 5, 7, 9, 2, 4, 6, 8!" Max had memorized this since he was little. As he let go of the propeller, the spring set it back into position.

"Make sure chocks are in before you throw!" yelled Grandpa.

Max knew a lot of aviation terms and what they meant. The chocks were the stoppers in front of the tires to keep the plane from rolling while it started up and "throw" meant to turn the propeller and stand back. Once it started, Max would step away, pick up the rope connecting the two chocks and pull them away from the wheels.

Max stepped up to the propeller once again, grabbed hold of it and with all his strength gave a mighty pull. The propeller turned, and the engine coughed and sputtered as it started to rotate. First with a stutter and then with a roar, the engine came alive. The propeller was spinning so fast you couldn't even see it.

With the engine roaring, as if begging to fly, Max climbed into the open cockpit. He put on his helmet and goggles, then buckled in.

"All set back there, Maxie?" Grandpa yelled out.

"Check!" Max yelled back. With that Grandpa pushed the throttle forward and guided Queen Bee out onto the field. Bumping along the grassy runway, Max always got butterfly feelings in his belly. Grandpa and Max were going flying!

Faster and faster they rolled across the field and gradually Queen Bee lifted into the air.

Houses and farms passed under them. They flew over grazing cows and horses running across the rolling hills. It was mesmerizing, like in a dream.

"

Bees can fly
two miles
from their hive,
pollinating more
than 1000 flowers
each day,
and find
their way
home.

Paul Stamets

Night Two

As the warm sun shone down on them and the wind blew across their faces, Max closed his eyes, and his imagination took over. In his imagination he wasn't just flying in Queen Bee, he was flying on the back of a real honey bee! A Queen Bee!

"Hello there," the bee said. "my name is Queen Bee, what is your name?"

Stuttering in disbelief Max replied, "Hello Queen Bee, mmm . . . my name is Max, umm, how are you?"

"Not too good Max and I think you know why. We need your help," Queen Bee explained as they flew over the apple orchards and farm lands. Max was nervous.

"No need to worry Max, we never hurt anyone, we have a big job to do caring for the flowers, fruits, and vegetables, but now we need your help."

"My help? What can I do?" Max said thinking there wasn't anything that could be done.

"I understand you listened to your science teacher Mr. Clark, right?" Queen Bee asked.

"Oh yes, it was fascinating what he taught us!"

"Then you understand that the pesticides used to spray the crops and the parasites that invade our hives are threatening our lives."

"Oh, yes, Max replied, "of course, I do understand. They know what is causing the problem, but nobody knows how to fix it."

That's what I am going to teach you, Max. Will you agree to help us?"

Max didn't know what to say. He couldn't believe that the bees wanted to teach him. *Why did they pick me?* He wondered.

"Yes, of course, I will help. I will do anything I can."

"That's all we need Max, is for you just to try. Let's begin - I have a lot to teach you. Hang on tight!"

They flew through the air with high speed and not in a straight line like Grandpa's plane. Darting and dodging through the air Queen Bee headed down to the ground, faster and faster and with the greatest of ease. With no effort, Queen Bee landed on a log lying on the ground.

"Wow!" Max yelled, "That was incredible!"

"Max, we don't have a lot of time. Let me explain. See this rotting log we are on? To you, it may look lifeless, but you need to understand that there is so much life living on this one log. See these mushrooms here?" She pointed to the mushroom growing on the side of the log and the one growing from the forest floor.

"This is the Reishi mushroom Max – this is one of three mushrooms the bees need to survive. We choose our homes very carefully, and try to make sure we are close to this and other mushrooms."

Max looked with amazement. "Mushrooms, what do you do with them?" Max asked.

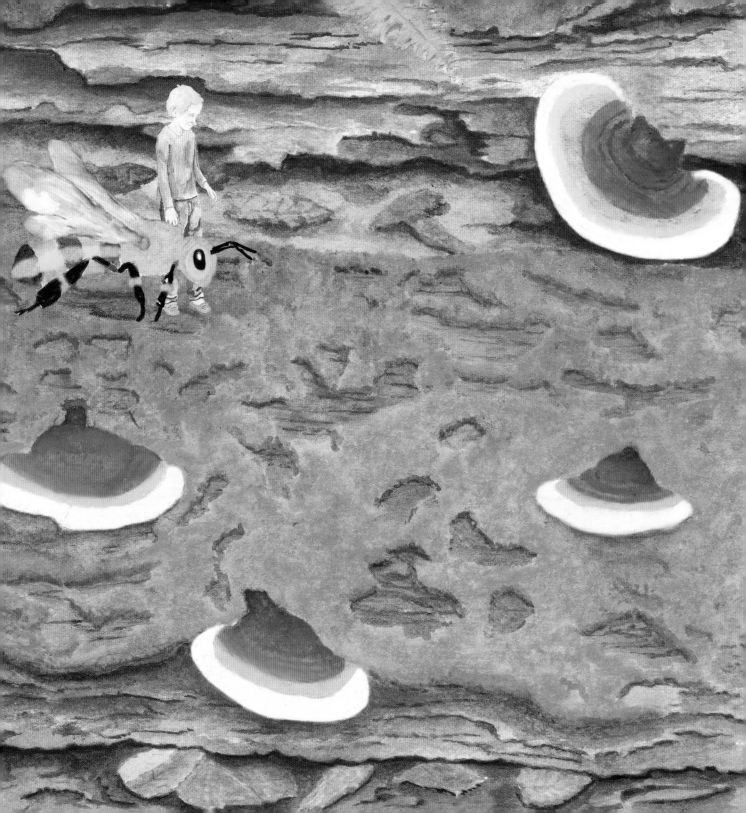

Queen Bee lifted up a piece of the rotten wood just under the mushroom, "See here? This white web looking stuff? It's called mycelium, it's the root system of mushrooms, and this is what helps keep all the bees healthy."

"There seems to be plenty of mycelium around, so why are the bees getting so sick?" Max asked.

When we drink from this mycelium, our immune systems stay very strong. It's how we can fly up to five miles looking for pollen, and it helps us fight off the disease-carrying parasites that invade our hives. The parasite is called the Varroa Mite, and it attaches itself to us and our brood spreading disease and sickness. These tiny mites infect us with viruses. The mites make us sick taking away our strength and energy so that we cannot fly back to the hive. And Max something very sad happens then. Nurse bees, seeing that we haven't made it back, must leave the hive to go gather the pollen - that means we cannot raise as many young bees as we need."

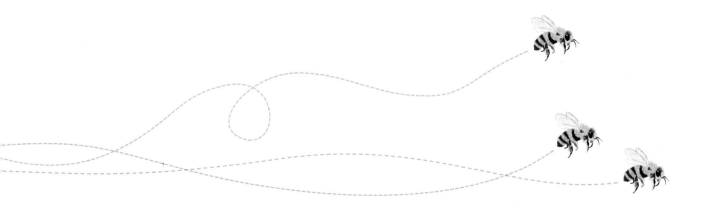

Max, the problems are also the things you can't see. Pesticides and chemicals are sprayed on the flowers and the crops to keep damaging insects away. The pesticides are also used to kill the weeds that grow. But those chemicals are washed down into the rotting wood and eventually into the soil as it pollutes everything it touches. Mushrooms are like nature's sponges – they suck up everything - the right nutrients the soil has to offer, and the chemicals that people put into the environment.

"The chemicals are very poisonous to us and very harmful. They weaken our immune systems. But that's not all Max, we lose our sense of direction, our sense of smell which makes it very difficult to fly to where the pollen is. So, you see, we're not strong enough to fly the great distances needed to collect pollen and then fly back to our hives.

"Now, this mycelium also grows just beneath the surface of the soil," said Queen Bee as she jumped down onto the ground and lifted up the soil. "As you can see Max, it is under here too."

"I had no idea, Queen Bee."

"I understand Max, most people don't."

"You said there were two other kinds of mushrooms you need, what is another one?" Max asked.

"Get on Max. I have something to show you."

Max held tight to the hair on Queen Bee's back, and she lifted into the air like a helicopter.

"Where are we headed?"

"To one of the oldest living species on earth." Queen Bee replied.

Excited to know which species it could be, Max asked, "Can you give me a hint?"

"Okay Max, I will give you a hint. They have been around since the glaciers melted, they are also home to many birds and wildlife, and they are critical to the bees survival. Can you guess?"

Well, this had Max thinking and without even trying to guess he said, "I have no idea."

"Max," Queen Bee began, "you will be surprised by what you don't know that you know."

As they flew Max noticed they were deep in the forest. "Where are we going? Why are we in the forest?"

As Max asked the question, they landed on a tall Birch tree. Max was amazed by the smoothness of the bark, not rough or jagged like most tree barks.

"This tree is beautiful Queen Bee. What is it?"

"It is the Birch Tree, Max. This tree gives everything it has back to the environment. Even when it dies, it still serves to support the wildlife of the forest."

"How can it do that?"

"Dead trees can remain standing for decades Max, even longer. A standing dead tree provides a great home for wildlife. Birds, insects and other animals make their homes in cracks and openings within these dead trees, while insects that feed on these dead trees provide the food for a larger web of forest life."

"Wow, I had no idea. With so much life on this tree, it doesn't seem right to call these trees dead." Max replied.

"As it dies, it has reached one of the most vibrant times in its life – serving a very beneficial role in the ecosystem. It is essential and one of the most active parts of the forest."

"Then why is it endangered, I mean, if it is so important why isn't anyone doing anything about it?"

Queen Bee turned and just looked at Max.

"Max, let me explain why we love the Birch, Beech, Elm, Ash and Aspen trees so much. See that dark black looking stuff growing there on the Birch tree," pointing to the side of the tree, "that is called the Chaga mushroom Max. We need that mushroom as much as we need the Reishi mushroom that you saw growing on the rotting wood."

"Why?" Max asked.

"Because, like Reishi" Queen Bee began, "they help us stay strong against infections, viruses, and the diseases that are caused by the mites."

"You said there were three mushrooms the bees need. What is the third?" Max asked.

Looking up the tree, Queen Bee notices another mushroom, the Amadou, growing on the side of the tree.

"For that, we need to fly up this tree a bit Max. Now, this mushroom will grow on other species such as Aspen, Maples, Poplars and Alder trees, but one of its favorite places to grow is right here on the Birch tree."

Hanging on to Queen Bee she effortlessly flew up the tree. "This mushroom is very unique," Queen Bee began. "It is called Amadou, and for thousands of years, it has been used by our ancestors for medicinal purposes. It loves to grow on trees throughout the temperate and boreal regions of Europe, and Asia and here in North America Max. It has even been found in Turkey and the Caribbean. It can be found on both dead and living trees."

As Max looked at the Amadou Mushroom, Queen Bee unexpectedly flew to another part of the forest.

"Now I am going to tell you something very few people know. Max, it is ancient knowledge, long forgotten. See those scratches there on the side of the tree? What do you think causes those?"

"Oh, I know, bears scratch the trees! They mark their territory by scratching tree trunks right?" Max answered, feeling proud of himself.

"Yes, that is right Max! But those scratches mean far more to us than you may know. Remember how I told you about mushrooms and mycelium growing on rotting wood? Well, when the bear scratches or bites the tree it creates an area of rotted wood and that allows the mycelium to take hold for mushrooms to grow. It also attracts insects which feed birds and other wildlife. The bees come and drink from the nectar of the mycelium Max, and this is very important because and it is this mushroom that helps our immune systems so we can live longer."

"Wow! That is amazing Queen Bee! I am so surprised by how many places mushrooms can grow. Can you explain why, if you have all these places where mushrooms grow, are the bees still in danger?" Max asked confused.

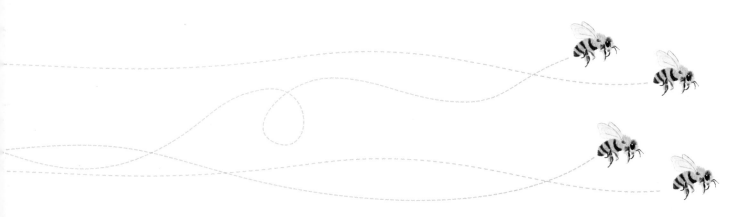

"You see Max, the truth is that besides the pesticides and chemicals, we face another serious danger as well - deforestation. The loss of forests to human activities, like clearing land and loss of old-growth forests, have taken away the sources of mycelium that we once had."

"Wait. Did you say paper companies?" Max asked confused, "Why paper companies? What part of this could they possibly play?"

"Max, we need the Reishi, Chaga and Amadou mushrooms to survive and so we need the sources those mushrooms grow on. When you start to reduce the places we can go for the mycelium and mushrooms that we need, then the health of all bees declines. Does that make sense?"

"Yes, of course," Max agreed, "but I still don't understand how paper companies play a role in all of this."

"To understand that is to know that timber companies clear-cut acres of forests and they cut the dead trees as fast as they find them. What happens Max, is that this creates not only a shortage of places where birds and other forest wildlife can live and gather food but reduces the options where these specific mushrooms can grow."

"The reason they cut dead trees is that they believe the dead wood increases the risk of forest fires. Loggers cut living and dead trees for sale to lumber and paper making companies. But those dead trees are not the cause of the fires Max, those fires are however part of nature's process. Dead trees are a vital part of the forest. People should appreciate them, along with the natural processes that create them, such as beetles and wildfires. Max, forests come in a variety of ways that also deserve protection. For example, trees with brown needles and trees with blackened bark.

This is called diversity. The same reason that the differences between human beings provide the essence of life, so it is with the forests Max, diversity is the basis of forest life.

"I never thought about it that way before. Everything is truly connected isn't it Queen Bee?"

"Yes Max, and that isn't the whole story. The bears are also in danger – timber companies are hiring hunters to shoot the bears to stop them from marking the trees that eventually create rotting or dead wood. But that dead tree, as we have seen, helps the whole forest, the wildlife and us Max."

"What? Why would they do that?" Max couldn't believe what Queen Bee was telling him.

As long as people keep thinking of nature as something they visit and not their home the struggles will continue. The timber companies log only live trees, Max, so they don't want the bears marking them. It's all a misunderstanding of forest life."

You have a chance to learn about the actual science of trees, of mushrooms and gain a greater understanding of nature. That's why we need you, Max. We need you to tell your friends and the other kids at your school about this and help create a better understanding – this will truly help the bees not only survive but thrive. You will help the trees and all the wildlife that depend on the forest for their lives. You can help the bears who also help the forest and they, in turn, help other wildlife. And Max, it is essential for the food source for your people.

Max, people need to understand that nature is intelligent. But do not be discouraged, you can communicate with nature and learn all you need to know so that you can help each other. The only reason people do not communicate with nature is that they have not decided to listen. There is an unexpected awareness of awe waiting for all of you. Max, we need you to help spread this knowledge to others. Do you see now why we need you, Max?"

With the breeze blowing, Max heard Grandpa say, "Max you awake back there? Look below!"

Max woke up suddenly, realizing that he fell asleep. He looked down at the ground in amazement, looking at the rolling hills and the trees below.

"Look, Grandpa, it looks like a giant puzzle."

"Yes, Max – it's all more interconnected than we realize."

Grandpa's words reminded him of his dream. As Max looked at the ground below, he thought about his dream and seeing life from the bee's perspective. Learning from Queen Bee, he now understood the challenges bees face every day.

Just then, the plane started to stall and stutter!

"Grandpa, what's wrong?" Max said starting to get scared.

"No worries Max, I have her under control! We just have to land her in an open field and see what is going on."

Max knew this wasn't good and he could hear it Grandpa's voice. He knew it well enough to know that Grandpa was just trying to keep calm and keep Max from panicking.

"

The majority of
modern medicines
originate in nature.
Although some mushrooms
have been used in therapies for
thousands of years,
We are still discovering new
potential medicines
hidden within them.

Paul Stamets

Night Three

Looking down seeing only the trees and hills, "Land? What land, where?" Max yelled back.

"No worries Max, you'd be surprised how little space old Queen Bee needs to land."

It was true, this plane could take off and land in very small spaces, and Grandpa found the perfect spot.

Landing safely right next to an apple orchard, Max thought about his dream again.

"Grandpa you're not going to believe this dream I had."

"Not now Max, we need to find out what's wrong with Queen Bee." Grandpa climbed out of his seat.

Lifting the engine lid, "Well, Max I see the problem. The airflow is being blocked to the engine – I can fix this, but I need some tools." Looking around for a nearby farm, Grandpa could see they had a long walk ahead of them.

As they began to walk, Max began, "Grandpa I had a dream while we were flying. I learned the secret life of bees. They taught me how to help fix Colony Collapse Disorder."

Grandpa started to laugh. "Max, you have quite the imagination," he said, as they made their way across the fields.

"No! Grandpa, listen!" Max said urgently, stopping in his tracks. "Grandpa, listen, I think I know how to save the bees!" And just as he said that, they came upon a rotten log with mushrooms growing on it. Just like the ones he saw in his dream!

Max explained about the Reishi mushroom and how bees rely on them. He lifted the topsoil up to show Grandpa the mycelium growing just under the surface, "This is mycelium Grandpa, and the bees drink from this root system to help them survive."

"Wow, Max, that is amazing," Grandpa replied curiously wanting to know more about Max's dream.

"Well, you won't believe it, Grandpa! The bees need the Birch, Beech and many other species of trees to help them survive, more than we ever knew. There are two other mushrooms too! The Chaga and the Amadou mushrooms."

"I know those mushrooms, Max because I took a mycology class in college. Did you know that mycology is the study of mushrooms and fungi? But nobody ever talked about this connection to bees."

"I don't think anyone knew about it Grandpa," Max replied.

As they walked across the fields and farms, Max told Grandpa all about his dream and what he learned. As Max described his dream, Grandpa got an idea.

"Come on Max let's hurry and get these tools so we can get back home. I have an idea, and we have work to do."

After getting the tools from the farmer, they fixed their plane and flew back to Grandpa's farm.

Get on the computer Max. Look up those mushrooms you were talking about and find out where we can buy some. Also, look up anything you can find on Colony Collapse Disorder."

Max started to research, and soon he found the university where they were doing studies on the effects mushrooms have on bees – he couldn't believe it. There it was, just as Queen Bee had taught him.

"Grandpa!" Max yelled, "I found it!"
Max showed Grandpa the Washington State University website. They described all the tests they have been conducting using the mycelium. They allowed the bees to have access to the Chaga, Reishi and Amadou mushrooms – it was working. The viruses were decreasing by 99%, and the life of the bees was increasing by 100%!

Grandpa called the university right away, "Hello, I would like to talk to the person in charge of the entomology department."

"Oh yes, the voice from the other end of the phone began, "that would be Dr. Sheppard. Please hold."

Grandpa explained his idea to Dr. Sheppard, "Perfect! Thank you so much." Grandpa said, and hung up the phone.

"What Grandpa?" Max asked excited, "What did he say?"

"Well, apparently several years ago, Paul Stamets was outside looking at his garden full of the Garden Giant Mushroom. He noticed the bees landing on the wood chips and digging down into them. He watched them for about a month. What he discovered struck him as odd. The bees came to the garden, dug through the wood chips and flew back to their hive. They seemed interested in only what was below the wood chips. It was through Mr. Stamets' observations that he discovered the bees sip nectar from the mycelium, little dew-like droplets oozing from the mycelium, just below the woodchips. Now some very smart scientists have been working on this bee challenge, and they do indeed have those mushrooms you talked about. Dr. Sheppard explained how this brilliant scientist, Paul Stamets, manufactures this liquid, he calls it mycelium extract, and they have a lot of it. He explained Mr. Stamets' commitment to planet and people, and that he has made this a priority. Mr. Stamets says that without bees every food manufacturer, every farmer, every store and all the people on earth will be affected.

I explained to Dr. Sheppard my idea, and he said it's worth trying. He agreed to send out gallons of the liquid mycelium extract right away."

"What idea Grandpa? What is your idea?" Max asked excitedly.

"You know how I used Queen Bee to spread those pesticides over the farmer's lands to help them fight the weeds and the insects from destroying their crops?"

"Yes," Max replied curiously.

"Well, once you explained to me about how we are causing harm to the bees and the environment, I started thinking, what if we replaced those pesticides and chemicals with the liquid mycelium you told me about, and we sprayed it everywhere? Can you imagine how that might help the bees?"

"Grandpa, that is brilliant! How do we do it?"

"Max, you're the one who had the dream about the bees, if you hadn't dreamt that I would never have thought about this idea."

Max was so proud of himself, he couldn't believe that he and his grandfather were going to help save the bees!

Come, Max, we need to make some minor adjustments on the sprayers and get Queen Bee ready for the liquid mycelium to arrive. It will be here tomorrow!"

Max and Grandpa worked all night getting Queen Bee ready. They swapped out the pesticide spray nozzles for nozzles Grandpa made in his workshop, and they bought new tanks to hold the mixture of water and mushroom extracts. Everything had to be perfect and clean.

As the sun came up over the farm, Queen Bee was ready.

Grandpa called all the neighboring farmers to explain to them their plan. He told them about the cause of colony collapse and why that may also be the reason for the delay in crop growth and what they can do to save the honey bees. The farmers all agreed to let them try.

As they were walking to the plane, Max started to take his usual position at the propeller, just as he did every time they went flying. "Max," Grandpa began, "wait, today is a special day and would not have happened if not for you. Today Max, you are taking control of Queen Bee, and you will fly her! Climb in Maxie, I'll take care of that propeller."

Max couldn't believe what he had heard. "Grandpa, are you serious? I get to fly Queen Bee!"

"Max, you deserve it – it's because of you that the bees will be saved and the crops will come back. If anyone deserves to fly Queen Bee today, it's you, Maxie."

"Okay Grandpa, ready?" Max called out.

"Ready captain!" Grandpa replied.

Max yelled, "CONTACT!" as he took hold of the stick.

Max and Grandpa flew all day until dark, spraying the farms and apple orchards. The news spread to towns and farms all over Ohio.

News trucks and reporters showed up to Grandpa's farm to interview Max and his grandfather. Soon the word was out. Everyone heard about Grandpa and Max and their idea to save the bees and the crops. Soon farmers from all over the state were asking for help. Other crop duster pilots heard the news and began to reconfigure their planes and sprayers to do the same as Queen Bee. Bee farmers helped too by filling the drinking feeders at the hives with liquid mycelium.

STAMETS HELPS SAVE THE BEES

Daily Bee

Citizen Scientists Mobilize to Save the Bees!

Paul Stamets and **Dr. Steve Sheppard** Team Up for Bees

honey*times*

THE **Bee** Guardian

MAX

THE BEE TIM

Boy and Grandfathe help save the Honey

The VARROA

Cause: Pesticides, Parasites and Pathogens

STORY OF THE YEAR

B

Schools across the Nation start BeeFriendly™ Clubs

Dr. Steve Sheppard Conducts Bee studies

WSU

BEE FRIENDLY!
THANK YOU FOR SAVING THE HONEY BEES!

Several weeks past. Just as expected, the farmers started reporting their crops were growing and they noticed a decrease in loss of honey bees – the bees were thriving!

The experiment worked!

As the last day of summer came, Grandpa drove Max to the bus station in Columbiana. "Well, Max, we had quite the adventure didn't we?"

"Grandpa, this was the best summer ever! I can't believe we helped save the honey bees. I can't wait to tell all my friends back home. They'll never believe me!"

"Oh Max, I have a feeling they just might have heard."

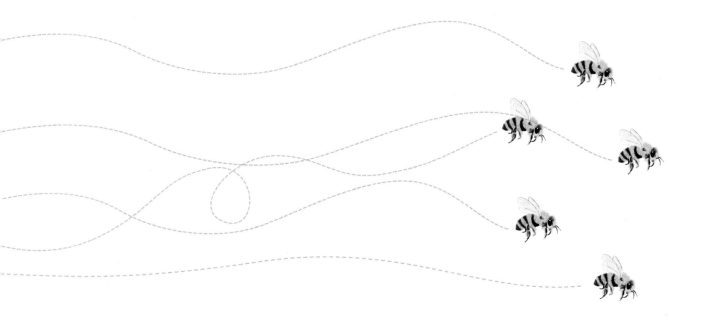

Sitting on the bus, Max looked out the window waving goodbye to his Grandpa. As Grandpa waved back, the bus pulled away from the station.

Soon Max was home, and there was his mother waiting for him.

"I'm so proud of you Max!" his mother said as she gave him a big hug.

"Oh, then you've heard?" Max replied wondering how she had heard.

"Did I hear about how you saved the honey bees? I should say so, everybody has heard about it! It's all over the news!"

Max had no idea that anybody at home had heard about what he and his grandfather had done.

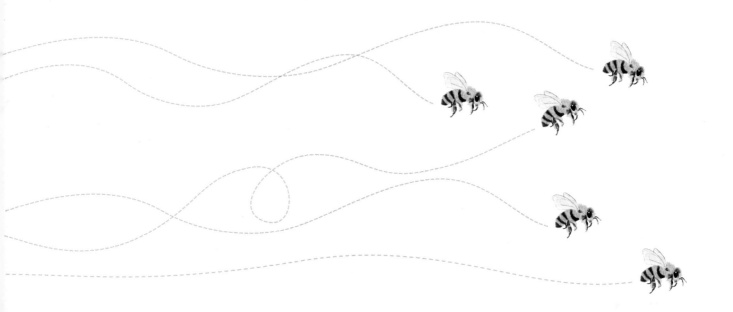

As Max walked into school, he couldn't believe his eyes. There was a giant banner hanging from the entrance of the school. It read,

"BEE FRIENDLY!
THANK YOU FOR SAVING THE HONEY BEE!

MAX sat at his desk, looking out the window thinking about his summer. He thought about the love he had for his grandfather and all that he had learned, from bears and trees to mushrooms and bees.

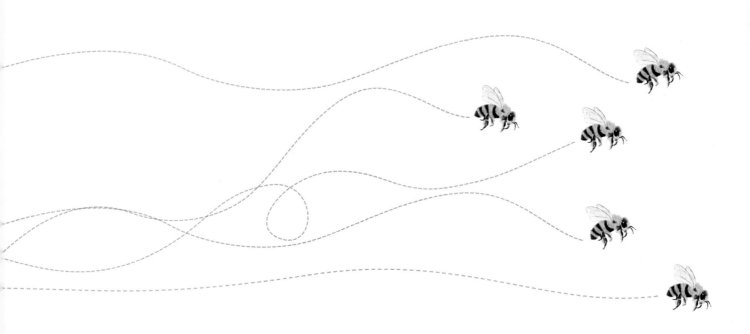

Notes from a Young Scientist:

Notes from a Young Scientist:

Bee Here Now

David Marshall, an award winning graphic designer, is a children's book author and illustrator. David's children's stories carry life giving, inspirational and meaningful messages (www.storyman.org). He believes that when kids read, they learn how and when to trust, dream and succeed. Storytelling helps children find a connection with their world, with their parents, teachers and with each other. Children face their challenges in many ways – his goal is to use stories to help children meet those challenges in life.

Photo: Trav Williams
Broken Banjo Photography

Paul Stamets is the author of 6 books, has discovered and named numerous new species of mushrooms and is the founder of Fungi Perfecti (www.fungi.com) and Host Defense Mushrooms. He has received many awards, including Invention Ambassador (2015) for the American Association for the Advancement of Science (AAAS), the National Award (2014) from the North American Mycological Association Award (NAMA) and the Gordon & Tina Wasson Award (2015) from the Mycological Society of America (MSA). An avid explorer into the old growth forests, his work focuses on the interconnections of mycelial networks and mushrooms within ecosystems.